CIRIA C745

Coastal and marine environmental pocket book (second edition)

Edited by
Sian John
Katharine Basford
Helen Craven
Nicola Meakins
Philip Charles
Sirio D'Aleo

Coastal and marine environmental pocket book (second edition)

John, S, Basford, K, Craven, H, Meakins, N, Charles, P, D'Aleo, S (eds)

CIRIA

CIRIA C745 RP1002 © CIRIA 2016 ISBN: 978-0-86017-750-0

First printed as C594 in 2003

British Library Cataloguing in Publication Data

A catalogue record is available for this book from the British Library

<table>
<tr><td colspan="3">Keywords
Coastal and marine, construction materials and products, construction process and management, construction resources and waste management, environmental management, regeneration and contaminated land, sustainability and the built environment, sustainable water management</td></tr>
<tr><td rowspan="5">Reader interest
Coastal and marine environment, construction materials, construction process, environmental management, impact avoidance, minimisation, mitigation, performance measures, planning, procurement, risk and value management, sustainability, waste management</td><td colspan="2">Classification</td></tr>
<tr><td>Availability</td><td>Unrestricted</td></tr>
<tr><td>Content</td><td>Advice/guidance</td></tr>
<tr><td>Status</td><td>Committee-guided</td></tr>
<tr><td>User</td><td>Site workers</td></tr>
</table>

Published by CIRIA, Griffin Court, 15 Long Lane, EC1A 9PN, UK

About this pocket book

Any construction project, irrespective of the location, size or nature of the development, will have environmental consequences. This is particularly true when working on the coast or at sea given the sensitive nature of the environment and the importance of this region for human activities. However, good management and construction processes can avoid or minimise impacts.

This pocket book focuses on the control of environmental impacts arising from construction practices in the coastal and marine environment (recognising that the coastal environment includes terrestrial elements).

It provides advice on carrying out construction works in coastal and marine environments while minimising impacts to these sensitive areas.

The following environmental issues (**Chapter 3**) are covered in the pocket book:

- biodiversity and geodiversity
- contamination and pollution prevention
- the historic environment
- nuisance
- resource management
- visual impacts.

The pocket book also highlights the issues site-based staff need to be aware of, and to plan for, when undertaking common construction processes (**Chapter 4**).

For more detailed guidance and further information readers should refer to CIRIA C744 (John *et al*, 2015).

Contents

1 Introduction

Construction works (see Figure 1.1) in the coastal and marine environment (eg ports, harbours, wind farms) will often be located within areas designated for their ecological, geological and/or geomorphological importance. So it is important to balance construction requirements with environmental protection.

Preventing or minimising the risk of impacts is a primary responsibility of all those involved in the construction industry.

At the site level there are increasing demands for high standards of environmental awareness through trained and experienced staff who can effectively manage the site and any issues that arise with the urgency they require.

Adopting good practice should reduce costs, reduce the risk of pollution, and can improve a company's image and performance, while contributing to sustainable construction.

Figure 1.1 ***Reinforcement of the current coastal protection, creation of new dunes and tidal and recreational nature at Waterdunen, the Netherlands (courtesy DEME nv)***

1.1 Working in the coastal and marine environment

The dynamic and changing forces of the sea set apart construction activities (see **Figure 1.2**) in the coastal and marine environment from similar activities on land. Waves, tides, currents and winds introduce risks and issues that need to be overcome by site staff to allow an operation to proceed successfully (eg vessels are subject to weather and wave limits).

Figure 1.2 Site works being carried out during calm weather conditions (courtesy Royal HaskoningDHV)

1.2 Consents and licences

The regulation of activities that can affect the environment vary across England, Wales, Scotland and Northern Ireland (see Useful contacts for details of UK regulators, environmental agencies, statutory nature conservation organisations (SNCOs) and heritage bodies).

A variety of consents may be required for works in the coastal and marine environment, and the requirement for them will depend on the activity proposed. Examples include:

- **Marine licence:** for activities involving the deposit or removal of a substance or object below mean high water springs (MHWS).
- **Protected species licence:** for permission to carry out an activity involving species protected by wildlife legislation.
- **Discharge licence:** for any activities involving water or sewage discharges, or dewatering of tidal excavation.

It is important to check consenting requirements with a relevant environmental advisor or manager before potentially disruptive activities are undertaken in sensitive areas.

1.3 General site rules

Adopting generic good practice measures is the first step to minimising potential impacts on any site. Successful environmental management also relies on effective communication. It is crucial to be aware of the key issues, have the relevant information to deal with them, understand the responsibilities and feedback to those in charge.

Good practice checklist: general site rules	
Find out who is responsible for the environmental management of the site	
Be aware of any consents and controls relating the works and the site (see Section 1.2)	
Ensure staff are aware of the sites environmental issues and sensitivities	
Follow good housekeeping practices	
Ensure that good site security is in place	
Make sure materials are safely and securely stored	

Good practice checklist: general site rules	
Follow endorsed pollution prevention practices	
Monitor the site and report any concerns	
Ensure staff are aware of the location of emergency equipment and who to contact on site in the event of an environmental incident	
Communicate good practice requirements to site workers, contractors and visitors	

1.4 Benefits

Adopting good practice from the outset of a project can lead to a range of environmental, social and economic benefits being realised:

Environmental

- **Reduced damage:** to the natural and human environment, often creating opportunities for enhancement (see Figure 1.3).
- **Resource efficiency:** by using less materials (eg concrete, quarried rock, sand and gravel).
- **Reduced disruption:** to local residents and businesses (ie visual nuisance, noise, dust, vibration).

Figure 1.3
Colonised artificial rock pool, Tywyn, Wales (courtesy Aberystwyth University)

Social

- **Reduced nuisance:** to neighbours by adopting best practicable means (BPM) to reduce noise, dust, vibration and light pollution.
- **Improved community engagement:** liaising with local communities before and during the project to keep them informed about works that could affect them.
- **Increased knowledge/skills:** for site-based staff learning and adopting good practices.
- **Increased local skills:** through employment (eg of apprentices) that can lead to longer term common growth.

Economic

- **Reduced fines:** the cost of dealing with breaches in legislation are increasing, and can be compounded by clean-up costs.
- **Use of fewer resources:** resulting in lower cost of materials.
- **Reduced cost of remedial activities:** cleaning up after pollution incidents or repairing damage to habitats can be expensive.
- **Reduced risks:** improving working practices and management on site will reduce the risk of loss of plant or staff down time, which subsequently reduce costs.

1.5 Obligations

Site staff should be aware of the legislative and corporate controls relevant to the project (ie those highlighted at site induction) and the good practice measures that should be adopted to meet them. This is often through a site specific Environmental Management Plan (EMP).

Adoption of appropriate good practice measures on site will:

- minimise noise and vibration
- prevent pollution of surrounding water bodies
- prevent disruption to protected species, habitats and neighbours
- avoid damage to protected habitats and species.

2 Preconstruction phase

This chapter identifies relevant site-based responsibilities before any works begin, provides guidance on the risks associated with prevailing weather and sea conditions, as well as on setting up and managing coastal sites.

> ***Plan ahead***
>
> *The earlier contractors can be brought into a project, the better. They can aid significantly in pre-construction planning and design phases, adding their knowledge and site experience to development of the scheme.*

2.1 The management framework

Team work is essential for the delivery of effective environmental management on site, including between the main contractor and their supply chain. To manage this process, the following steps should be followed:

1 **Clearly establish the project's environmental obligations**

- This will help to define roles and responsibilities for the site (refer to **Step 3**):
- **Steps 1 and 2** can be achieved through:
  - review of relevant documentation (eg Environmental Statements (ES), Environmental Action Plans (EAP) etc)
  - discussions with environmental representatives, the local authority, relevant environmental agency or SNCO/heritage body
  - seeking guidance from specialists (eg an ecologist).

2 **Identify the environmental hazard and sensitivities particular to the site**

- Become familiar with all aspects of the site – especially those identified as containing protected species and/or habitats.
- Sensitivities or risks may require special construction procedures, temporary works or specifically trained staff – this should be communicated (ie by using site boards as shown in Figure 2.1).
- Use staff experience of similar sites.

3 **Establish an environmental management structure and plan**

- Define the environmental responsibilities and authority of staff on site (including those involved in monitoring).
- Details of staff with defined roles and responsibilities on site should be recorded in an EMP, alongside procedures to manage potential issues on site.
- It is crucial that the EMP is accessible, revisited when appropriate and communicated to all site-based staff.

4 **Train staff**

- Appropriate training of staff and a clear definition of responsibilities will help to minimise the potential for accidents to occur.
- Training should include site inductions and toolbox talks, and supplemented by targeted training activities.
- It is important that training is revisited throughout the project to take account of issues arising and to ensure that staff are familiar with and carrying out current good practice.

5 Monitor actions and effects

- Monitoring is likely to be required where significant potential impacts have been identified during the project planning.
- It may be carried out by regulators, but can also be undertaken by the developer, and agent, or contractor.
- Results of monitoring undertaken should be recorded, which can serve to confirm that procedures are being followed or where there are opportunities for improvement.

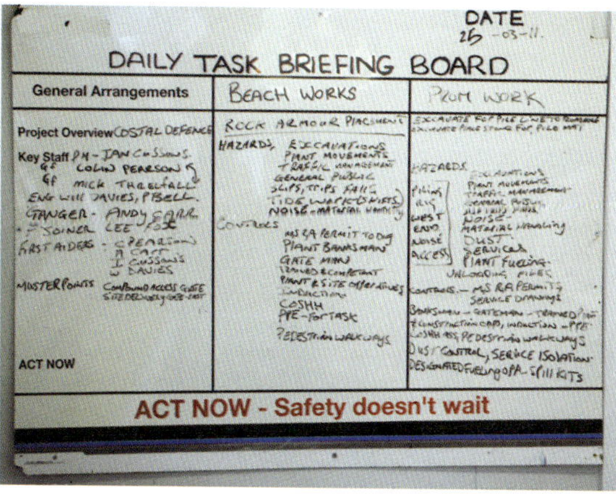

Figure 2.1 Site daily task briefing board, Colwyn Bay, Wales (courtesy VolkerStevin)

2.2 Setting up and managing the site

2.2.1 Stakeholder management

Timely, ongoing and clear engagement with stakeholders, including regulatory agencies, conservation bodies, the public, site neighbours and other stakeholders (eg fishermen and coastal managers) is very important when setting up a site, and should include:

- confirming existing contacts and contact details
- communication of an agreed engagement plan to maintain relationships with partner organisations and the public
- identifying and assigning responsibility to appropriate site staff to undertake the necessary liaison during the construction phase
- ensuring that stakeholders are kept up-to-date with the construction programme.

Figure 2.2 **Eastoke Point stakeholder engagement event (courtesy Eastern Solent Coastal Partnership)**

Good practice checklist: liaising with the local community	
Where relevant, hold an event in a local venue (see Figure 2.2) to explain the construction process, programme, potential impacts and future benefits taking into consideration language and cultural needs	
Identify and regularly update key local community representatives regarding the progress of the works	
Liaise with the following stakeholder to inform them of works and expected timing of activities that may cause disruption: • harbourmaster • local fishing industry and their representatives • local boating and sailing clubs • recreational anglers • any other interested stakeholders.	
Send out Notices to Mariners	
For large-scale projects such as wind farm developments consider arranging educational visits to local schools	

2.2.2 Site management

Sites should be tidy, secure, have clear access routes, take account of the effects of weather events and be well signposted. Collectively these measures encourage efficiency and promote site safety.

The vast majority of environmental accidents or causes of complaints stem from one or more of the following:

- adverse weather conditions at sea
- faulty equipment/leaks
- ignorance
- negligence
- carelessness

- vandalism
- theft.

When planning a site layout, offices and equipment should be sited to minimise visual intrusion. In coastal areas (as elsewhere), consideration should be given to minimising any visual effects on neighbours, recreational users and marine users.

Attention should also be paid to site security. Contractors will be liable for environmental damage caused by vandals (eg tipping out liquids from drum and containers, opening taps on tanks containing fuel). Measures to avoid damage as well as the risk of theft should be considered when planning site layout (eg use of fencing as shown in Figure 2.3) and implemented during construction.

***Figure 2.3** Fencing around site (courtesy Royal HaskoningDHV)*

Good practice checklist: housekeeping	
Segregate waste as it is produced and remove frequently (or reuse/recycle as appropriate)	
Stockpiles of material (eg sand) should be covered, bound or fenced to ensure that material is secure	
Ensure that skips have lids to prevent their contents becoming airborne and that they are emptied before they become overfilled	
Keep the site tidy and clean (storing and locking away appropriate equipment and materials at the end of each day)	
Ensure material and plant storage areas are properly managed (see Figure 2.4)	
Repair and repaint site hoardings when necessary and remove fly posting	
Frequently brush clean wheel washing facilities	
Do not leave plant unattended in public areas or near to the intertidal zone	
Ensure materials are not stacked against the site boundary or fence	
Move valuable items from public view and store these in locked containers	
Ensure that materials potentially hazardous to the environment are secured and that fuel outlets are locked	

Figure 2.4 ***Double bunded generator (courtesy BAM Nuttall)***

2.2.3 Management of materials

Improving the management of materials on site reduces waste and increases site efficiency. Environmental benefits associated with the efficient use of resources include minimising the amount of waste sent for disposal, with the additional advantage of saving money.

Good management includes efficient ordering, appropriate storage and effective waste management. Where site staff follow established procedures for managing materials there will be fewer incidents of spillage and contamination arising from incorrect storage or handling, and less damage to materials and components.

It is good practice to develop and implement a Resource Management Plan (RMP)/Site Waste Management Plan (SWMP) from the outset of a project to ensure that waste is managed correctly throughout the works.

Good practice checklist: material ordering and delivery	
Ensure that the right quantity and quality of materials is ordered	
Plan delivery of materials to arrive at the time when they are needed, this reduces the length of time they have to be stored on site and, subsequently reduces the potential for damage, pollution and theft to occur	
If ordering low volumes, consider the issue of higher transport requirements and the potential nuisance that could result	
When ordering, find out what form the materials will be delivered in, so that appropriate unloading plant can be arranged	
Ensure deliveries are received by a member of site staff who is able to supervise the delivery, carry out a quality inspection and ensure materials are unloaded to the appropriate place	

Good practice checklist: material storage	
Ensure supplier instructions on storage and delivery are followed	
Provide a larger area for material storage on site to limit disruption by adverse sea conditions and maintain continuity of construction activities	
Store valuable materials in secure areas to reduce risk of theft	
Store materials in a central area of site away from watercourses and protected species and/or habitats – particularly where materials are potentially polluting (ie fuels oils)	
Store materials away from waste storage containers and from vehicle movements that could cause accidental damage	
Ensure liquid chemicals (eg fuels, acids, coolants, oils) are stored in appropriately (portable or permanent) bunded facilities that will prevent accidental spillage, leakage or jetting of liquids into the environment	
Secure lightweight materials to protect them from wind damage or loss	
Ensure materials are labelled appropriately and stored in suitable containers	
Ensure that when storing materials the effects of extreme weather in the coastal zone and at sea is considered and appropriate action taken, for example by securing them in a locked container or anchoring them to the ground	
Make sure that appropriate emergency response equipment is located near to the stored material and that staff know how to use it (see **Section 2.2.6**)	

Good practice checklist: material handling	
For liquid chemical products and raw materials (eg paints, thinners, solvents), ensure supplier instructions on handling are followed	
Undertake an assessment of materials stored on site that require careful handling (eg Control of Substances Hazardous to Health (COSHH) materials). Ensure appropriate equipment to handle them correctly is available on site and site-based staff are trained to handle them	
Ensure appropriate dust controls (eg water carts, sprinklers or sprays) are employed to reduce dust emissions during ground disturbance activities	
Minimise distance of free handling of liquid materials as far as possible, ie avoid moving drums or containers without appropriate controls	
Use drip trays and ensure appropriate spill response equipment is at hand when handling liquid chemicals	
Designate refuelling areas and enforce restrictions to ensure that such activities are only undertaken in these areas	
Refuelling areas should be designed to contain spills and contaminants and comprise of: • hardstanding • water treatment system (where appropriate)	
Before moving materials at sea ensure weather conditions have been considered	

2.2.4 Traffic management

An organised site with well-defined traffic management can leave a positive impression on local residents. Without appropriate management of site traffic, adverse environmental impacts can include:

- delays to local traffic
- safety hazards and severance both on and off site
- direct damage to the environment.

Implementing good practice can help ensure people living and working near the site are not unnecessarily affected by emissions, noise and the visual intrusion of queuing vehicles.

Vessel traffic and deliveries by sea should also be properly managed as these can potentially affect other businesses and marine users (ie ports, fisheries or recreation areas). The relevant authorities, such as the Harbour Authority or Coastguard, should be consulted and weather and sea conditions monitored.

Good practice checklist: managing site traffic	
Ensure all delivery drivers are aware (ie through a site induction) of traffic restrictions at and around the site.	
Good signage should be provided to guide drivers to and around the site, with access control onto site where appropriate	
Schedule deliveries to avoid the peak traffic periods (eg commuting or 'school run') to minimise potential disturbance	
Arrange delivery timings to avoid queuing outside the site boundary	
Instruct drivers to switch off engines when vehicles are waiting	
Put in place on-site speed limits to reduce the risk of dust or collisions	
Load and unload vehicles within the site boundary	
Wheel cleaning on site and road sweeping should be carried out to keep the public highway clear of mud and debris	
Plan parking for site staff vehicles	
Consider getting regular site vehicles to display identification	
Ensure that deliveries by sea are carried out in compliance with what has been agreed with the relevant authorities	
Provide a larger area for material storage on site to limit disruption by adverse sea conditions and maintain continuity of construction activities	

2.2.5 Incident response planning

It is essential that pollution prevention strategies are used when working in the coastal and marine environment.

Emergency response plans to deal with potential pollution incidents need to be in place before setting up a site (see Figure 2.5).

Consideration needs to be given to controlling and containing solid and liquid pollutants (eg keeping absorbent powders to remove larger spills on-board vessels) along with the preparation of detailed action plans for dealing with emergencies, eg for unexploded ordnance (UXO) (see Section 3.4.5).

Good practice pollution prevention includes ensuring that:

- risks to biosecurity have been identified
- all waste water produced on site is disposed of appropriately and cannot enter controlled waters
- appropriate containers are available for storing and transporting waste
- plant and equipment is serviced and managed to minimise air pollution
- specialist equipment to clean up oil is kept on site and on-board vessels
- lighting is correctly set up to avoid navigation hazard and pollution
- all liquids are appropriately stored to prevent spillage.

Stop

Switch off any sources of ignition if safe to do so and try to identify the source of the pollution eg stop the flow of a liquid by righting containers or closing taps or valves.

Contain

Attempt to stop pollution spreading by using items from the specified spill kit:

- dam the flow with absorbent materials, booms, earth, or sand.
- plug drainage channels/grills with drain covers, even if some of the spillage has already entered the drainage system.
- divert flows away from drains and watercourse.

Notify

Notify the foreman/supervisor immediately with the following information:

- location of where the substance has entered the drain/watercourse or is affecting the environment
- substance involved
- quantity of substance involved
- reason for incident.

Inform environmental regulator if substance has entered a watercourse or drain

Contact Tel: 0800 80 70 60

Figure 2.5 Example incident response procedure

2.2.6 Spill kits

Spill kits consist of equipment to contain and absorb spills on land and at sea and need to be suitable for the particular spillage(s) likely to occur.

At the start of the project it is important to assess the number required and proposed locations for the deployment of kits, for quick access across the site or vessel (ie near to where they may be needed).

The content of a spill kit (see Figure 2.6) will depend on the project, but is likely to include:

- absorbent granules
- string
- floating 'booms' or 'sausages'
- PPE, ie gloves, goggles and overalls
- absorbent mats
- drain covers
- polythene sheeting and bags.

Figure 2.6 Spill kit station (courtesy BAM Nuttall)

2.2.7 Managing monitoring equipment

To meet project obligations (and as good practice) monitoring of environmental conditions within and/or beyond the construction site (eg noise, marine mammals) may be required.

As such, it may be necessary for a contractor to maintain monitoring and mitigation equipment on site, such as noise meters, water samplers and chemical test kits. Note that in some cases, spare equipment need to be available on site as well.

To remain effective, equipment will need to be properly serviced and inspected, with attention paid to transport and storage of the equipment when working offshore.

2.2.8 Estimating and planning for sea conditions

Coastal and marine sites can be affected by adverse weather and sea conditions (eg wind, waves, currents and tides).

The impact of these conditions should be considered by the contractor before works start on site, and developed into a plan that includes information on the operation and a stability assessment for plant addressing issues relating to weather and the sea state.

The plan should consider the following:

- the amount of time land- or marine-based plant are likely to be unavailable for work
- excessive wave action causing motion that make it impractical to operate plant
- whether it is appropriate to completely shut down construction activity for the winter and/or to avoid disturbance to overwintering birds
- whether it will be necessary to design temporary works to withstand predicted extreme events.

All site staff should be aware of the plan and of the measures to undertake when the weather and sea conditions are adverse.

2.2.9 Use of plant in intertidal conditions

A variety of land-based plant can be used in the intertidal area (eg crawler cranes, excavators, see Figure 2.7). However, determining which plant should be used will be dependent on the substrate and habitat present at the site, and the exerted ground pressure of the plant, especially over soft ground and on an incoming tide.

Figure 2.7 Rock armouring works, Kirkcaldy, Scotland (courtesy VolkerStevin)

Restrictions on or requirements for the use of particular plant in the intertidal area may be in place to avoid impacts on the environment in designated sites. Access for plant across sensitive habitats needs to be planned carefully to minimise impacts as well as its size and travel speed.

3 Construction phase

This chapter highlights the key environmental challenges that can arise during construction and provides guidance on minimising and mitigating potential impacts.

3.1 Biodiversity and geodiversity

3.1.1 Designated sites

A number of designated sites occur in the coastal and marine environment, with large areas of coastline and the seabed protected for the habitats and species found there.

Designated sites should be protected from construction activities as far as possible. Signs and fences, such as chestnut paling mounted on a scaffold framework and, where appropriate, flags and buoys, should be used to identify the protected areas.

Site staff should be aware of the restrictions and conditions to work on the area and abide by their requirements. Such requirements would be included in the site-specific EMP.

Failure to comply with the protection given to designated sites can result in legal action and significant fines, as well as the requirement to restore the habitat to its previous state or replace the habitat (which can be very expensive).

3.1.2 Coastal and marine habitats

Coastal and marine habitats (see Figure 3.1) are highly complex, dynamic environments. They can quickly change in response to natural influences, wave exposure and climatic drivers as well as be significantly affected by human activities.

Figure 3.1 Silverdale saltmarsh, Lancashire, UK (courtesy R Saunders)

Factors that can cause damage to coastal and marine habitats include:

- water quality changes including turbidity, temperature, contaminant levels, pH, salinity etc
- physical disturbance of (including removal/ disruption of substrate and deposition) materials on the seabed
- de-watering
- changes to currents and to erosion and/or deposition patterns
- disturbance, eg noise due to marine works or presence of people or machinery
- development of the coastal margin.

The potential impacts of construction activities can be reduced by applying good practice on site, but where impacts do occur, they will be most visible through the following:

- discoloration, retreat of and visible damage to habitats

- sick, injured or dead birds and/or fish close to the site
- crushed plants, visible signs of trampling and vehicular damage to flora.

If any of these signs are apparent or it is suspected that a protected plant or animal may have been affected by operations, stop work and seek advice before continuing.

Actions to take to avoid impacts include:

- explaining the environmental importance of the site to staff and clearly setting out approaches to good practice relevant to the site
- ensuring staff are aware of where protected habitats occur and that they know of any specialist treatments/ methods to be used in or near to such areas
- installing posters identifying risks and good practice measures (see **Figure 3.2**)
- adhering to an EMP.

Figure 3.2
Sign showing environmental sensitivities of a site (courtesy BAM Nuttall)

3.1.3 Fish and shellfish

Construction activities can affect fish and shellfish through:

- changes in water quality
- loss of habitat
- noise and vibration
- pollution (including by light).

Should damage occur, costly remedial works to mitigate the effects construction activities have caused may be required. There could also be a requirement for compensation if a fisherman's livelihood is shown to be affected because of any construction activity. This could be in the form of:

- interruptions to fish migration
- reduced access to commercial fishing areas
- reduced catches due to construction.

The protection of fish and shellfish can be achieved through the adoption of good practice, and where appropriate look for opportunities to create or enhance habitats (see Figure 3.3).

Figure 3.3 Artificial reef balls installed at the C-Power offshore wind farm to stimulate the marine environment (courtesy DEME nv)

It is fundamental to raise site staff awareness of what could be found on site (see actions in **Section 3.1.2**) and to pay particular attention to noise levels (see **Section 3.4.3**) and to the risk of pollution incidents (see **Section 3.2.2 and Section 3.2.4**).

Regular inspections should be made for visible signs of fish distress and/or mortality. If identified, seek advice before continuing.

3.1.4 Marine mammals, basking sharks and turtles

Around the coast of the UK, 11 species of cetacean (whales, dolphins and porpoises) commonly occur, as well as two native species of seal (the grey seal and the harbour (or common) seal).

In UK waters all species of marine mammals are protected, and as such certain activities may be restricted due to protection in place.

Marine mammals are particularly sensitive to human-made noise that can lead to disturbance, injury and even death. Potential noise sources include:

- construction noise, eg such pile driving
- the use of explosives
- military sonar
- seismic and bathymetric surveys (eg using a multi-beam echo sounder)
- vessel noise associated with various activities
- dredging and aggregate extraction
- installation of cables and pipelines.

Regular site inspections should be carried out to look for signs of distress or mortality. If found on site or if it is suspected that an animal may have been affected by operations (ie if they display unusual or abnormal behaviour, signs of distress), stop work and seek advice.

Seals

Grey and harbour seals are found around the coast of the UK. They can be observed at haul-out sites in their greatest numbers during their breeding season (grey seals in autumn, harbour seals (see Figure 3.4) in June and July) and annual moult (January and February for grey seals, August for harbour seal).

Figure 3.4 **Harbour seal** *(courtesy Royal HaskoningDHV)*

Seals of both species can be sensitive to disturbance at haul-out sites due to approaches by vessels and walkers. Seals are also sensitive to underwater noise, but do not rely on it for finding food, so are considered less sensitive than cetacean species.

In addition to impacts from noise, seals may be injured or killed via the use of ducted propellers.

Basking sharks

The basking shark (see Figure 3.5) is the largest fish (10 m to 11 m in length) occurring in UK waters (between April and September). They are commonly around western Scotland, the central Irish Sea and south western England.

Figure 3.5 Basking shark (courtesy G Skomal)

They have been assessed as having a very high risk of extinction in the wild and are protected from harassment and deliberate killing. They are also affected from those activities producing noise and vibration (eg pile driving, rock blasting or dredging).

Turtles

Turtles are commonly found off the west coast of the UK between August and October. All species are protected, but it is not an offence to help turtles if entangled or stranded, or to temporarily hold dead turtles for later examination by experts.

Recognising problems and taking action

- Careful planning and effective management of the construction site should ensure that injury and damage to species and habitats is minimised.

The following actions should be taken:

- be aware of the species that potentially could be encountered on site
- ensure that posters and signs showing pictures of the species of concern, identifying risks and goods practice measures are installed on site
- adhere to any Marine Mammal Mitigation Protocols (MMMP) and codes of practice.

3.1.5 Birds

All birds and their nests are protected – with some receiving a higher level of protection than others.

Birds often use construction sites as breeding areas, including nesting on scaffolding or machinery. If this occurs the equipment cannot be used until the birds have finished nesting and any young have left the nest. The area may also need to be sealed off to prevent any disturbance.

Every effort should be made to prevent disturbing large aggregations of birds, especially at high tide in coastal environments, as birds during this time have a limited area on which they can roost and/or feed.

Site inspections should be carried out regularly and works should be undertaken outside of the breeding season (if possible), and in compliance with the permission sought from the relevant SNCO.

Be aware that there may also be restrictions in place in terms of heights of plant permitted for use near to overwintering birds to avoid visual disturbance.

3.1.6 Invasive species

Invasive non-native plant and animal species are one of the greatest threats to biodiversity worldwide.

Once established, it is extremely difficult to get rid of an invasive species in the marine environment, making it very important to implement measures to prevent the species arriving in the first place. For example:

- pictures of key potential invasive species should be provided on site to ensure that early detection is possible
- thorough inspection of all clothing and equipment should be undertaken
- hosing down or pressure washing of equipment on site should occur
- use disinfectants (approved for use by the environmental regulator) to clean plant and equipment after use in the water
- if washing facilities are not available, equipment should be carefully contained, eg in plastic bags, until it can be washed
- clothing and equipment should be thoroughly dried (48 hours before reuse) – boots and nets should be hung up to dry.

These steps are in-line with the 'Check-Clean-Dry' campaign to reduce the spread of invasive species such as the killer shrimp or Zebra mussel (see **Figure 3.6**).

Figure 3.6 Zebra mussel (courtesy Paul Beckwith, BWW)

The environmental regulator advises that a biosecurity/ non-native risk assessment is carried out before starting any marine or coastal activities.

This is especially where vessels are being used and/or materials are being brought on to site. Also, check that good practice biosecurity guidelines for prevention of spread of invasive non-native species (INNS) are followed.

Further information on identifying INNS are provided by:

- GB non-native species secretariat (NNSS): **www.nonnativespecies.org**
- Marine Life Information Network (MarLIN): **www.marlin.ac.uk**

3.2 Contamination and pollution prevention

3.2.1 Water

Water quality is a key factor in determining the health of the marine environment and fisheries (eg fish farms). It can also have an impact on a range of additional factors, such as tourism and recreation in coastal areas.

Once a pollutant enters the sea or a watercourse it is extremely difficult to remove or control, so it is crucial that steps are taken to prevent water pollution.

There are many materials, wastes or by-products that arise from construction activities that may affect coastal and marine water bodies, including:

- silts and clays re-suspended by the works
- cement
- oil and chemicals (ie from plant machinery and processes)
- other wastes, such as woods, plastics, sewerage and building rubble
- effluent from washing/cleaning operations
- surface water run-off
- re-suspended heavy metals, bacteria and viruses.

If an incident occurs, in addition to clean-up costs, fines and legal fees, the overall costs of a water pollution incident will include disruption to the construction activities and non-productive time for the equipment and site staff.

Measures for avoiding incidents are set out in the following checklists.

Good practice checklist: concrete, cement and bentonite	
Take particular care when producing, transporting and placing these materials, especially if working next to the shoreline or a watercourse	
Use methods to minimise grout loss during shutter pours	
Place covers over freshly poured concrete to prevent the surface washing away in heavy rain	
Do not hose down spills into surface water drains	
Carry out washout in a designated impermeable and contained area. Do not allow washout water to flow or drain into the sea or any drains	
Reuse washout water as much as possible	
Dispose of washout water by tankering off-site or discharge to a foul sewer in accordance with the required agreement in place	

Good practice checklist: water quality	
Where appropriate, ensure plans are in place to appropriately manage surface water and to treat it as required	
Establish and mark the location of any surface water drains and foul sewers on-site	
Deploy turbidity curtains to contain suspended solids within a designated area (see Figure 3.7)	

Good practice checklist: storing and using fuels and oils	
Locate storage away from the site drainage system and the shoreline. If this is not possible, ensure adequate measures are identified to prevent or contain any spillage (eg blocking drainage points)	
Keep storage areas away from vehicle access routes to reduce the potential for collisions	
Storage must be sited on an impermeable base within a bund designed to contain at least 110 per cent of the volume of the largest tank or 25 per cent of the total storage capacity, whichever is the greater	
Securely store all ancillary equipment (eg valves, hoses) within the bund when not in use	
Ensure that tanks are correctly labelled as to their contents and capacities	
Keep a store of spill response equipment at the fuel facility and bowsers	
Protect storage facilities against vandalism and theft with the facility being locked off when not in use	
Regularly inspect containers and tanks and ensure these are appropriate for the material stored in them	
Use drip trays under all static plant (eg pumps and generators), particularly during refuelling from mobile plant, and empty them regularly	
Establish a refuelling procedure on site and appoint refuelling staff on site. Where appropriate, contact the harbour authority and avoid times when vessels are passing to avoid any swell/unexpected wave action	
Use fuel and oil dispensing systems as appropriate to the location of the project	

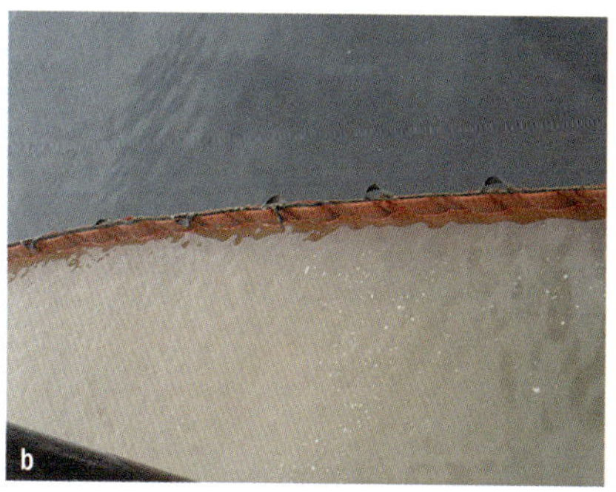

Figure 3.7 Turbidity curtain before deployment (a) and deployed (b) (courtesy BAM Nuttall)

3.2.2 Sediment

Generally, contamination results from the previous uses of a site (eg industrial use, waste disposal and ship yards).

Contaminated sediments are likely to be present at sewerage or other outfalls, where boat maintenance has been undertaken or where previous pollution incidents have occurred.

When contaminated sediments are found on site, these should be managed with extra care as there is the risk of waterborne contaminants being released outside of the site and consequent harm to humans, flora and fauna.

Sediments should not be allowed to spread and should be recovered and disposed of in a manner agreed with the appropriate environmental agency.

Problems can arise when:

- handling or dredging contaminated sediment or soils
- excavating on previously industrial land
- contamination is encountered or exposed (eg rupturing historically disposed barrels)
- contaminated dust arising from excavation, loading of lorries and transportation is windblown.

Contamination can be spread (or further contamination caused) due to:

- windblown dust
- stockpiling contaminated material on clean ground without adequate protection
- stockpiling materials containing contaminants liable to leach out to underlying ground or nearby water bodies, such as saline lagoons, without adequate protection
- slippage or discarding of contaminated land into water

- spillage of contaminants such as oil onto the ground or into water during construction
- de-watering that draws in contaminated groundwater from nearby sites
- discharge of contaminated de-watering output into nearby water bodies
- disposal, re-suspension and spillage of contaminated sediment from dredging operations.

Good practice checklist: avoiding causing or spreading contamination	
Do not stockpile contaminated soil unless it cannot be avoided	
Stockpile only on hardstanding and a liquid-tight area away from the sea or watercourses	
Never stockpile contaminated material on unprotected clean soil	
Contain drainage run-off and dispose of appropriately to avoid discharge to water bodies	
Cover stockpiles to prevent windblown dust or to minimise the ingress of rainwater or sea spray	
Be especially vigilant when working close to or within tidal limits regarding presence of contaminated material, spillage of materials from plant and possible instability	
Prevent the spread of contaminated dust	
Be careful when handling, storing and using oils and chemicals – provide bunding/secondary containment for liquids	
Ensure plant have appropriate apparatus (eg shrouded head or enclosed grab) to prevent potential contaminants from being spread during removal	

Good practice checklist: avoiding causing or spreading contamination	
Ensure measures in place to contain contaminants following removal from sea bed	
Check whether a contingency plan for unforeseen discovery of contamination is in place	

3.2.3 Land

Some incidences of contamination (eg large spill of a pollutant occurs) will be easy to spot, while for others it will be necessary to look for a range of tell-tale signs during site activities.

The following may indicate a contamination incident has occurred:

- discoloured soil (eg chemical residues)
- fibrous texture to the soil (eg asbestos)
- presence of foreign objects (eg chemical/ oil containers)
- evidence of previous soil workings
- evidence of underground structure and tanks
- existence of waste pits
- made ground (ie artificial ground where ground level is raised by man's activities and not due to a natural cause)
- old drain runs and contamination within buildings, tanks, flues etc
- release of fumes and smells (such as petrol, oils, bad eggs)
- sediment close to outfalls or previous pollution incidents
- sediments that leave an oily sheen or iridescence when in contact with water
- oily sheen on water surfaces may indicate leaching of contaminants from surrounding soils

- material behind retaining walls and dock walls where contaminated backfill may have been used or liquid contaminants may have been trapped.

> ***Emergency response***
>
> ***If contamination is suspected (identified by strong smell, oily sheen etc) do the following:***
>
> - *stop work immediately*
> - *seal off the area*
> - *report the discovery to the site manager*
> - *as far as possible identify the extent and cause of contamination, eg spillage on site*
> - *attempt to contain contaminants*
> - *seek expert advice.*

3.2.4 Oil spill response

It is essential that an effective oil spill response plan is in place to deal with any spills as quickly and efficiently as possible to minimise the potential impacts.

Spill kits should be stored in marked bag(s) or container(s) in well sign-posted location(s). They should be stored close to where they are likely to be needed.

It is recommended that tamper-evident seals be fitted on the containers so that it is clear when equipment has been used. Regular checks should be made to ensure that all of the equipment is present, undamaged and that spares are available as required.

Further details on incident response planning and spill kits can be found in Sections 2.2.5 and 2.2.6 respectively.

3.3 Historic environment

The historic environment in the UK (on land, on the coast and at sea) is an extremely rich source of knowledge about the past, whether or not it has been afforded statutory protection. The historic environment includes heritage assets such as:

- submerged and intertidal prehistoric remains, eg shipwrecks and aircraft wrecks, and Palaeo-landscape features
- archaeology and built heritage, eg foreshore fish traps and ports and harbours, and historic shoreline management structures
- legacy of coastal military defences
- historic landscape and seascape character.

Many heritage assets are protected by legislation and these must be clearly marked (either physically, or on maps or navigation software) before any construction or excavation works start. Also, to accommodate known heritage features that are undesignated by legislation, site staff must be similarly informed of their spatial position and extent.

Should the potential for buried archaeology exist in the development area, then appropriate plans and procedures should already be in place, which will help manage unexpected discoveries.

However, all site staff should be made aware of unexpected archaeological finds, which could be associated to deeper and more extensive remains. Materials to look out for during excavations include:

- shipwreck or aircraft wreck material (see Figure 3.8), eg worked timber and metals
- burned or blackened material
- brick or tile fragments
- coins
- pottery
- bone fragments (human and animal remains)

- skeletons
- timber joints or post holes
- brick or stone foundations
- infilled ditches.

Figure 3.8 Recovered aircraft engine (Junkers 88) discovered during dredging in outer Thames Estuary (courtesy Wessex Archaeology © London Gateway Port)

At any time during construction, archaeological remains could be found unexpectedly. The steps set out in the good practice checklist should be followed.

Good practice checklist: discovery of archaeological remains	
Inform all site staff spatial position and extent of known heritage assets (designated and undesignated), within and close to the development area	
Inform all site staff that potential archaeological findings may be found on site	

Good practice checklist: discovery of archaeological remains	
Stop work immediately if any artefacts or historical features are identified on site	
Mark the area to avoid further disturbance	
Contact the site manager who will inform the relevant authorities to seek advice on the next steps	
Should works be allowed to re-start, adhere to the specific working methods to avoid disturbance to the remains. Adhere to expert advice from relevant authorities	

3.4 Nuisance

3.4.1 Dust

Dust emissions can arise during land preparation, construction activities on a site and from sediments on deck.

Dust may cause nuisance to neighbours, which includes exacerbating existing health problems, as well as soiling of building surfaces, vehicles, gardens and washing.

Dust blowing onto water bodies and coastal habitats (such as saline lagoons and saltmarsh) can damage the ecology too, particularly if the receptor is sensitive to particulates.

On a site, dust can cause mechanical or electrical faults to equipment and can increase the abrasion of moving parts and clog filters and air conditioning systems.

The potential also exists for the reputation of a company to be affected and legal action to be taken, with cost and programme implications.

Dust generated can be reduced by following the good practice measures here.

Good practice checklist: avoiding dust generation	
Haul roads	
Select suitable haul routes away from sensitive areas	
Regularly sweep paved access roads and public roads using a vacuum sweeper, but avoid dry sweeping of large areas	
Limit vehicle speeds – the slower the vehicles the less dust will be generated	
Ensure vehicles entering and leaving sites are covered to prevent the escape of materials during transport	
Damp down, where appropriate. For long-term sites, consider the installing wheel washing facilities	
Demolition	
Use enclosed conveyors and chutes for dropping demolition materials that have the potential to cause dust	
Regularly dampen the chutes	
Locate crushing plant for materials away from sensitive sites	
Beach recharge	
Notify local residents of proposed activity and avoid rainbow spraying when the wind direction is towards sensitive receptors	
Plant	
Clean wheels of vehicles leaving the site to avoid spreading mud onto roads	
Stick to the site speed limits to reduce the risk of dust clouds	
Ensure load tipping is carried out with care and with minimal drop heights	
Earthworks and excavations	
Re-vegetate or seal temporary or completed earthworks as soon as possible	
Keep earthworks damp where practicable	

Good practice checklist: avoiding dust generation	
Consider potential dust emissions when planning work phasing	
Materials handling and storage	
Locate stockpiles out of the wind to minimise the potential for dust generation	
Keep the stockpiles to the minimum practicable height and use gentle slopes	
Compact and bind stockpile surfaces, and re-vegetate long-term stockpile	
Minimise the storage time for materials on site	
Store materials away from the site boundary and downwind of sensitive areas	
Ensure all dust-generating materials transported to and from site are covered	
Minimise the height of fall of materials	
Avoid spillage and clean up spills as soon as possible	
Damp down, where appropriate	

3.4.2 Emissions and odours

The main concern with regard to odour is its potential to cause an effect resulting in annoyance or nuisance to surrounding neighbours.

It is important to prevent emissions and odours, as far as practicable, both to protect workers and the public and because they may affect the environment.

Processes involving the use of fuels and the heating and drying of materials commonly emit fumes, odours or smoke.

Good practice checklist: preventing emissions and odours	
Vehicles and plant	
Keep site vehicles and plant well maintained and regularly serviced	
Control deliveries to site and staff car parking, to minimise queuing	
Ensure engines are switched off when they are not in use	
Keep refuelling areas away from the public and ensure storage systems are fully sealed and spillages are minimised	
No fires on-site	
Do not burn waste materials/tyres on site – it is illegal	
Waste storage	
Use covered containers for organic waste and site rubbish and empty frequently	
Remove organic waste (eg weeds and other vegetation) before it begins to decompose	
Chemicals on site	
Store fuels, chemicals and other hazardous substances appropriately	
Ensure storage systems are fully sealed and spillages are minimised (and handled appropriately)	
Take account of the prevailing wind conditions when arranging activities that are likely to emit aerosols, fumes, odours and smoke	
Position temporary site toilets away from public areas	
Equipment	
Use ozone-friendly refrigerants	
Perform regular leak tests	
Recover gases during servicing and at the end of the equipment's life	
Use water based or low-volatile organic compound (VOC) paints	

3.4.3 Noise

Noise can have wide ranging effects during construction on the coast and in marine areas. These relate to:

- hazards to workers or annoyance to neighbours
- injury and disturbance to marine mammals
- disturbance to marine and coastal birds
- injury and disturbance to fish and other aquatic life.

If complaints are received and investigated by the relevant local authority or conservation agency, this will often lead to delays to the construction programme and subsequent cost increases.

The most likely source of problems is piling, which is generally the loudest activity on a construction site.

Other activities, such as excavation and plant and generator use, may provide a source of complaint, particularly at night.

Figure 3.9 Noise shroud on piling rig (courtesy VolkerStevin)

Good practice checklist: working near to sensitive receptors	
Be aware of and adhere to the restrictions or conditions that are a requirement of the consent	
Reduce noise to minimum practicable levels	
Schedule noisy activities to avoid sleep disturbance	
Adopt silenced piling techniques	
Use noise shrouds around piling equipment where feasible (see Figure 3.9)	
Place acoustic fencing in front of receptors where it is not possible to place it close to the source of the noise	
Place stationary noise sources (eg generators) as far as possible from sensitive receptors	
Fit construction plant, such as generators and pumps, with appropriate silencers and acoustic enclosures and/or hoods (where appropriate) in order to minimise the effects of noise	
Employ marine mammal observers (see Figure 3.10) to monitor for marine mammals and to communicate with on-site staff to take appropriate action when identified	
Ensure site staff are using the most appropriate equipment and method to minimise disturbance to marine species (eg fish and mammals)	
Slowly increase noise levels to encourage species to move away from the area before significant noise emissions	
Send letters to local residences, providing them with details of the works	
Publish timings of noisy works in the local newspaper, and provide details of works on posters or on information notice boards on site	

Figure 3.10 Marine mammal observer at work, Copland Dock, Orkney (courtesy BAM Nuttall)

3.4.4 Vibration

At high levels and over long periods of time, vibration from construction activities (ie piling, demolition, haulage) can cause damage to:

- buildings
- nearby structures
- sensitive equipment, eg computers.

Damage is more likely to occur:

- if there are structures very close to the source of the vibration
- where ground conditions are such that the transmission of vibration is high
- where water levels are high allowing vibration to travel further.

Where potential damage is possible, it is recommended that site boundary ground borne vibration levels are monitored before and during the relevant activities, to provide evidence of actual effects. Vibration is defined as nuisance, so is subject to a legal process if complaints are made and upheld.

Good practice checklist: reduction of vibration levels	
Use alternative methodologies to percussive piling or driving machinery	
Use machinery/equipment creating high as opposed to low frequency noise and vibration	
Use dense, heavy bases rather than light bases to minimise induced vibration levels	
Undertake the works during low tide when groundwater levels are lower	

3.4.5 Unexploded ordnance (UXO)

Ordnance that was once deployed during military activities but failed to initiate as intended or which, following the cessation of armed conflict, has been dumped or otherwise discarded/lost at sea, can present a threat to the marine environment.

The principal issue is that some human activities, such as those listed here, can lead to UXO detonation:

- trawling
- dredging
- installation, maintenance or decommissioning of marine infrastructure
- coastal defences
- ports and harbours
- cables and pipelines

- marine renewable energy farms
- oil and gas facilities.

The consequences of unexpected detonation of UXO can include:

- causing death or serious injuries to site staff and marine fauna
- destruction of vessels, plant or infrastructure
- destruction or damage to marine habitats or other assets, eg marine archaeology.

Emergency response

If potential UXO are found on site:

- *stop work immediately*
- *mark out the area to avoid further disturbance*
- *report the discovery to the site manager*
- *seek expert advice.*

3.5 Resource management

3.5.1 Waste

Waste is defined as 'any substance or object which the holder discards, or intends or is required to discard' (by the Waste Framework Directive, WFD).

By following good practice pollution risk can be reduced, money saved and prosecution avoided.

When considering materials used on site as waste, see the waste hierarchy, which will help save money and minimise the amount of waste generated (see Figure 3.11).

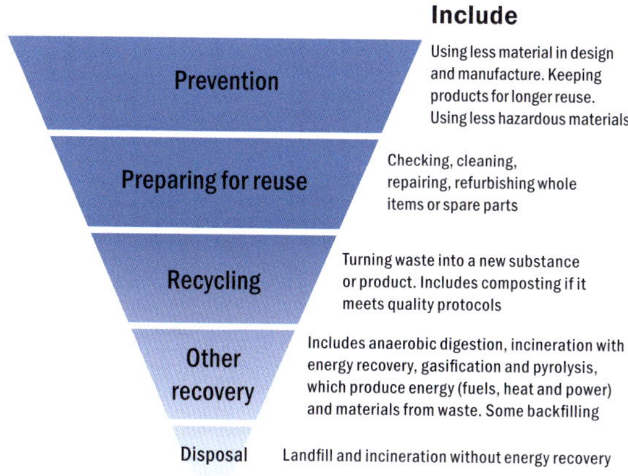

Figure 3.11 The waste hierarchy (courtesy WRAP)

Good practice checklist: storage of waste	
Segregate waste types as they are generated to maximise opportunities for recycling or reuse	
Mark waste containers clearly with their intended contents – consider colour coding (see Figure 3.12)	
Use containers suitable for their contents and regularly check their condition	
Minimise the risk of accidental spillage or leaks by siting containers in impermeable bunded areas	
Ensure that wastes cannot be blown or washed away and are stored securely to prevent theft or damage	
Store highly flammable liquid waste in an appropriate container	

Figure 3.12 Segregated skips (courtesy BAM Nuttall)

3.5.2 Dredged materials

Excavated dredged material is defined as waste.

Due to the salinity of the material arising from the marine environment, there can be restrictions on its potential for use in other marine and coastal projects.

Examples of the beneficial use of dredged material include the protection and creation of salt marsh and mud flats, which can provide useful flood and coastal defence.

However, the type of material being recovered will determine its potential recovery use, as some materials will be better suited for beneficial use than others.

3.5.3 Energy and water efficiency

There are legislative and corporate drivers in place to reduce energy consumption and CO_2 emissions as well as water consumption at the site level. Reductions can occur by installing water and energy meters on site to record usage, and put in place measures to reduce it, eg water saving toilets and taps.

Many actions are common sense, as they go hand in hand with financial savings achieved through the reduction of energy use. Energy efficiency measures can include:

- consideration of transport impacts:
  - deliveries to site
  - use of vehicles on site
  - how workers travel to and from the site
- managing on-site energy use, eg site office heating, cooling and lighting
- ensuring plant and machinery is:
  - not left running when not in use
  - well maintained and operated efficiently
  - fuel efficient.

Water consumption can be reduced through a combination of technological and behavioural change, which includes consideration of the water hierarchy (see **Figure 3.13**):

- installation of water efficient fittings (such as dual flush toilets)
- installing rainwater harvesting systems for non-drinking water
- reporting and fixing leaks and/or drips
- installing triggers on hoses.

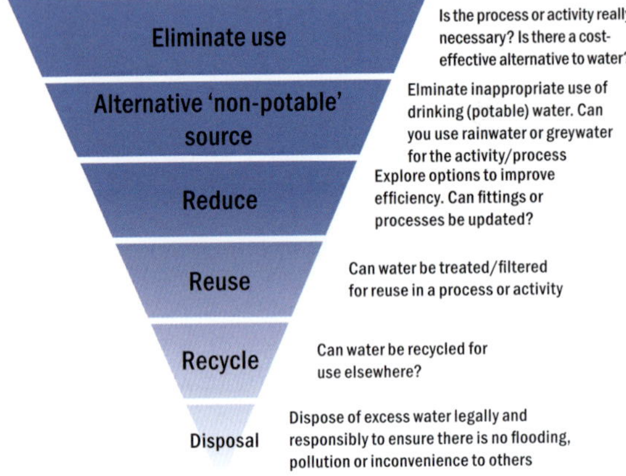

Figure 3.13 The water hierarchy (courtesy Green Construction Board)

3.6 Visual impacts

The potential impact of a project on the surrounding landscape and seascape is an important consideration for all projects in the marine and coastal environment.

The significance of any visual impact will depend on the nature of the surrounding environment, especially where the site is located in or within the visible range of a designated, protected or officially recognised important landscape or site of nesting for overwintering birds.

When setting up a site the proposed mitigation measures to reduce the sea/landscape or visual impact incorporated into the design will need to be followed through the construction process.

See **Section 2.2.2** on setting up and managing a site to minimise visual impacts.

4 Construction activities

This chapter describes the good practice measures that should be adopted for 20 of the most commonly undertaken construction activities (presented alphabetically).

4.1 Concrete pours and aftercare

- Control the storage, handling and disposal of shutter oils.
- Do not spray shutter oils close to or over watercourses.
- Should a spill occur ensure that an appropriate spill response procedure is in place and carried out (see Section 2.2.5).
- Use oils that are biodegradable and environmentally friendly where possible.
- Consider use of suction opposed to blowing dust and debris out of formwork to avoid nuisance, eg dust.
- Consider use of washout system to neutralise discharge from concrete placed in the inter-tidal zone.
- Use screens and/or muffled equipment to minimise noise generated from vibrators.
- Control placing of concrete to minimise risk of cement leaking into the watercourse.
- Consider generated waste and plan their handling and disposal, eg curing compounds.
- Dispose of wastewaters appropriately as they may be contaminated with sediments and cleaning materials.
- Do not allow washout water to enter any watercourse or groundwater.
- Surplus concrete should be avoided where possible. Excess should be stockpiled, recycled and reused by crushing and screening.

- Ensure batching plant has dust suppression system installed and a high level alarm to prevent overfilling.
- Carry out pre-use checks on ready-mix concrete wagons to check for defects, eg hydraulic hoses.
- Underwater concreting should be undertaken with care to minimise potential of polluting surrounding water environment.

4.2 Demobilisation (post-construction)

- Dispose of waste generated appropriately.
- Ensure that the proper documents are handed over to clients (eg approved licenses and consents and records of compliance with these, monitoring results, records of community consultation).
- Ensure the seabed is cleared of all debris – with appropriate surveying needed.
- Remove construction plant and site offices from site.
- Reinstate habitat/vegetation where it has been removed or relocated.

4.3 Demolition

- Avoid noisy sorting operations on site or at high physical elevation.
- Consider localised screening of operations or, subject to other constraints, carry out sorting operations at lower elevations (eg on the foreshore) or off site.
- When working tides, try to organise noisy operation for day shifts (note that times of low tides may not permit this).
- Monitor noisy activities to ensure that no limits set out in any consents are exceeded and that activities on the site are not a nuisance to neighbours.

- Stockpiled material should be suitably covered or dampened or bound to prevent dust arising.
- Be aware of potential archaeological discoveries during demolition activities (ie remains and artefacts) and contact the appropriate specialists immediately should any finds be discovered.
- When operating a crusher ensure it is located away from sensitive receptors.
- Ensure the discharge from crushers and screens onto conveyors is enclosed as far as practicable to minimise the effects of noise and dust.
- Ensure road transport of screened/crushed material is sheeted/in covered wagons.
- Ensure effective use of crushed and screened materials on site.
- Dispose of any materials in accordance with the relevant duty of care, segregating waste and disposing of any hazardous material appropriately.
- Underwater demolition needs to be controlled to minimise sediment mobilisation and ensure that all demolition material is removed from the seabed.
- Consider the noise/vibration impacts on sensitive species such as fish and marine mammals when undertaking underwater demolition.

4.4 Dredging (including disposal of materials at sea)

- Where sediment disturbance needs to be avoided, use an appropriate technique to minimise the disturbance of sediment.
- If an area is populated by protected fish species or marine mammals, it may be necessary to consider methods such as bubble curtains (in low flow

conditions) or acoustic screens to keep fish clear of the working area.

- Certain pathways may need action or monitoring to allow the passage of fish.
- Sediment plumes can be managed by working on certain tidal states, and can also be minimised through the optimisation of:
  - trailing/cutting speeds
  - pump/suction discharge rates
  - limiting overflow
  - reducing water intake and using return flow
  - shielding the cutter suction head.
- Odour problems may arise from the hoppers and from the sediments as they de-water.
- Monitor noise levels to demonstrate that noise is kept to agreed acceptable levels, and to minimise impacts on sensitive receptors (birds, people etc).
- Have noise mitigation measures in place when working with hard substrates or carrying out activities that cause vibration (eg mechanical de-watering, compaction).
- Be aware of the potential for the presence of UXO in the area of work.
- Respect the protection of any geological features or important habitats identified.
- Observe and adhere to traffic routing measures and restricted areas at all times, particularly in or near to habitats or features noted for their environmental fragility or sensitivity.
- Ensure implementation of any appropriate measures or protocol to identify and record, and if necessary preserve features and archaeological relics.

4.5 Drilling and blasting underwater

- Check and dispose of any unexploded charges used as part of blasting operations.
- Consider as part of blasting operations the following:
  - use of a protection shield, eg an air induced bubble curtain
  - use of rubber mats to prevent flying debris
  - monitoring of fish movements and appropriate response to minimise any impacts.

4.6 Excavation above water level

- When water does enter excavations, take measures to avoid it becoming contaminated and dispose of it appropriately.
- If asbestos is uncovered unexpectedly during digging operations halt operations at once and refill the excavation.
  - asbestos is hazardous and must be managed according to regulations
  - exposure of asbestos to the open air can result in widespread contamination as particles are easily airborne
  - remove site staff immediately and secure the area
  - contact site management immediately.
- Plant and vehicles used to transport materials may cause nuisance, eg mud and noise.
- Use a wheel wash to minimise dirt on public highway.
- Spoil arising from excavation can be recycled if not contaminated (subject to any applicable waste exemption/permits):
  - crush any rock arising and use on or off site
  - use to form noise bunds and for landscaping.

- Be aware of the potential for the presence of UXO in the area of work.
- Be aware of potential archaeological discoveries during excavation activities of undisturbed ground, eg remains (see **Section 3.3** for further advice).

4.7 Enabling works (platforms and foundations)

- Remember to plan ahead by considering the need for permissions:
  - licences to construct (eg platforms) or to dispose of dredged material
  - permits to discharge (if applicable)
  - access roads across flood defences may need require a license.
- Check whether there are any sensitive habitats that may be affected by the works that may need to be avoided.
- Ensure works have been surveyed for and cleared of any potential UXO.
- Identify working conditions required for installation activities.
- Pay particular attention to the need to undertake mitigation, such as 'soft starts' to pile driving operations.
- If there is any uncertainty or if a particular issue has not been covered, contact client and/or the local authority.

4.8 Geotextiles

- Use of geotextiles may reduce washout of fines from new bund or revetment, meaning less material is required while construction is stronger.
- Put in place procedures to manage waste generated:
  - rolls and central cardboard tube
  - off-cuts and ends of rolls.

- Ensure that any geotextiles being used in the works are secured at the end of each shift/day so that they are not washed away by the tides or storms.

4.9 Grouting (and tremie concreting)

- Protect sensitive areas from dust generated by blowback resulting from blockages or overfilling from pressure grouting by working within an enclosure.
- Prevent release of polluted water in excavations that has been displaced by grouting.
- Prevent release of cements and bentonite slurries – use a settlement tank to remove sediments and have in place permissions to discharge the effluent.
- Deal with slurry waste and waste additives appropriately.
- Locate batching areas carefully to minimise risk of causing nuisance, eg dust.
- Put in place a cleaning operation to remove silt or debris to prepare the bed or bottom of a cofferdam, pile for grouting.
- Discharge of excess grout into water and unnecessary waste should be avoided when grouting joints and gaps in existing structures, eg masonry.
- Ongoing monitoring of surrounding surface waters may be required due to the risk of grout seeping/loss.

4.10 Installation of pipelines and cables (including cable burial and protection)

- Work should be timed so as to avoid sensitive seasons for wildlife and tourism etc. Careful consideration should be given to mitigating any impacts on the local community.
- Ensure site staff are aware of the potential impact of locating or moving materials and plant over sensitive habitats, eg sand dunes.

- Ensure appropriate signage and barriers are visible, eg barriers are in place to communicate that an area should not be entered.
- Where avoidance of habitats cannot be achieved, consider the use of protective matting or raised track-ways to prevent damage.
- Ensure that the public are safely excluded from works that may dissect the beach and prevent access along it, eg watersports.
- If any archaeological remains are found, work must be stopped immediately and the archaeological specialist contacted.

4.11 Land-based plant (including maintenance)

- Prevent surface water run-off of pollutants, eg oils resulting from maintenance by using drip trays.
- Ensure appropriate trained staff carry out repairs to plant at regular intervals.
- Develop a protocol for disposing of wastes from maintenance, eg oil.
- Encourage use of biodegradable oils and recycling of used oils.
- Put in place appropriate plans to recycle tyres.
- All fuel must be stored in bunded tanks:
    - with a 110 per cent capacity of the material being stored (eg oil drum)
    - designate an area a suitable distance away from the water's edge for refuelling of vehicles
    - surface refuelling area and bund as surface water run-off may be contaminated.
- Prepare for spillages

- display site emergency response procedures at the plant maintenance area
- ensure a spill kit is kept there and that all site staff know how to use it
- carry a spill kit in all repair vehicles.

- Be aware of tidal and weather conditions on the coast, especially when working within the limits of wave and water action.
- Store all plant and equipment appropriately and secure during storm conditions to prevent damage to equipment.
- Land-based plant and workforce can affect the natural environment through direct disturbance of habitats and wildlife.
- Plant and trucks can damage natural features through crushing of sediment, plants and geological features.
- Do not encroach into sensitive areas marked out and keep to selected access routes and agreed working areas.
- Thoroughly clean plant that has been working in the marine environment before use elsewhere to prevent the spread of invasive/harmful species.

4.12 Marine vessels and plant (including maintenance)

- When refuelling vessels at sea have plans in place to contain any spillage on the vessel for subsequent removal to shore when possible.
- Follow International Maritime Organisation (IMO) guidelines when dealing with ballast water (IMO, 2012).
- Carry out a biosecurity/non-native risk assessment when antifouling to avoid the transfer of invasive species for vessel operation.
- Take care when the hose is returned after bunkering to

avoid any surplus fuel being discharged directly into the water.

- When operating, avoid collisions as any subsequent emissions, eg oil may have significant impact on wildlife, habitats, and fisheries.
- Ensure that all fuel tanks are safely secured to remove the risk of damage (eg via collision) that could result in leakage (ie of oil) into the water.
- Consider how anchor wires and ropes affect nearby businesses and marine users (ie fishing activities or navigation channels) and inform them. Cables can be 100 m to 150 m long, so the inaccessible area might be considerably large.
- Use modern fall-pipe vessels to tranship rock or implement measures to avoid rock spillages, as these can affect the seabed and produce unwanted noise.
- Spilt rock on the seabed should be recovered as soon as possible.
- Regularly maintain and service all marine plant to avoid oil and/or fuel leaks into the water.
- Ensure protective measures are in place in maintenance areas to reduce risk of spills, eg oil.
- Consider having 'marine skirts' available on site to contain any spills, eg oil.

4.13 Masonry

- Avoid spillage of grout into the water by ensuring:
  - all joints are adequately sealed and pointed before grouting
  - grouting is started at relatively low pressures to avoid blowing joints.

4.14 Nourishment and reclamation

- Grading, filling and compacting operations, eg for beaches, can generate unacceptable levels of noise, so consider possible screening of activities or timing to avoid more sensitive periods.
- Ensure areas used by the public are fenced off, with appropriate signage, to prevent accidents.
- Consider monitoring parameters such as noise and vibration, and dust to demonstrate compliance with legislative or contractual requirements.
- Be aware of the potential for increased suspended sediment to smother nearby habitats or shellfish beds, and take appropriate measures to avoid any impacts:
  - discharging only on particular states of the tide
  - construction of internal bunds
  - positioning and direction of the outflow pipe
  - avoiding discharge in adverse conditions
  - consider monitoring suspended solids at sensitive locations during works.
- Early and ongoing consultation and effective mitigation may prevent unnecessary damage, eg through screening, notice boards, newsletters.
- Ensure that any fill being used is not contaminated and avoid pollution of waters, wildlife, habitats, fisheries and natural features by sediment from filling.
- Protect surrounding land or aquatic habitats/species from the leakage or spillage of nourishment materials, eg using bunds or brushwood fences.
- Ensure appropriate safety precautions are taken such as fencing and signage to keep the public out of areas at risk of sand to liquefying.

4.15 Painting

- Surface preparation:
  - it is essential to make proper provision for recovery of old paint flakes during shot blasting
  - it should be carried out in a controlled manner, under sheeted enclosures that allows the material removed to be adequately collected and disposed of.
- Material removed should be checked for contaminants and dealt with as hazardous or special waste if necessary.
- Shot blasting operations should:
  - be carried out behind screens to prevent excess nuisance (ie, noise, dust)
  - consider undertaking dampening down to contain dust.
- Turbidity curtains should be used to contain suspended solids within a defined area.
- Site staff involved in the paint operation should be:
  - suitably competent and trained
  - provided with the correct PPE, eg breathing apparatus.

4.16 Piling

- Generally there are two options to install piles in a marine environment, which are influenced by the sea state/ weather conditions:
  - floating platform
  - fixed platform.
- Planning and/or design issues that may need to be considered early include:
  - the merits of steel versus timber piling, and solutions rather than piling
  - when to place sealant on to sheet piles
  - when to pile in sensitive areas.

- The use of floating platforms is restricted and is not suitable for:
  - drilled piles, except in very calm conditions
  - raking piles, except in very calm conditions
  - works that require the ability to resist torque
  - works close to navigational channels.
- For bored piling/rock socketing, removal of waste ('spoil') and flushings is the significant environmental impact to be considered and needs be managed appropriately.
- *In situ* concrete piles will require the same considerations during construction as set out for other *in situ* concrete structures.
- Seek advice on the paint/coating to be used, if applicable, to ensure that components are non-hazardous to marine life.
- Piles can be ordered with factory applied coatings. The quality of the finish and robustness of the paint system is enhanced because of the controlled conditions.
- If any archaeological remains are found, work must be stopped immediately and an archaeological specialist contacted.

4.17 Rock works and placement of concrete units

- Take necessary measures to prevent wildlife, habitats (including fisheries), natural features and archaeological heritage being affected by delivery of rock to site by sea.
- Deliveries can often be quickly offloaded during one high tide by discharging at the edge of the jetty and on the following low tide an excavator retrieves the rock for placing in the works.
- Where transhipping of rock is not possible it may be

necessary to construct a temporary offloading facility that can be serviced at all states of the tide (although this may be impractical and may need much cleaning up afterwards).

- Be aware of dust plumes caused by delivery of rock to shore and take appropriate action to minimise impacts, such as avoiding onshore wind, or consider spraying with water.
- Dust on rock armour can cause plumes of discoloration, which can be an aesthetic issue as well as an impact on marine or terrestrial ecology.
- Use biodegradable and environmentally-friendly shutter oils to mitigate any emissions.
- Offloading of rock can be extremely noisy, where possible, limit offloading to less sensitive times of the day.
- If any archaeological remains are found, work must be stopped immediately and the archaeological specialist contacted.
- Considerable land areas may be required close to the site for the production, curing and storage of precast armour units.
- Creation of suitably flat temporary areas can have:
    - a considerable impact on coastal/ marine features (eg beaches/dunes)
    - consequential impacts on wildlife and habitats and there may be associated safety issues.
- Ensure suppliers inspect quarried material before delivery and look for failed ordnances. If wires and charges are found on site, the area must be cleared and the local police advised.
- Ensure public access to the beach/coastline is controlled by fencing and appropriate signage, and that safe access is provided elsewhere if possible.
- Waste can occasionally arise during placing of precast

units, when units become damaged during placing – remove and recycle appropriately.

- Remove and appropriately dispose of materials that tend to blow about, eg shutters.

4.18 Site clearance (pre-construction)

- Follow established procedures if contaminated land is discovered (see Section 3.2.3).
- If protected species, eg wildlife or vegetation, are identified, seek expert advice to implement protection measures or relocate these.
- If invasive species are present, seek expert advice.
- Archaeological finds (including human remains) should be treated appropriately.
- Take measures to avoid pollution of site water supply or nearby watercourses.

4.19 Temporary works

- Take appropriate action to minimise the risk of pollution when carrying out temporary works near to or on watercourses.
- Consider restricting on site fabrication, eg of steelwork, to less sensitive times and providing noise barriers to minimise impact.
- Reuse materials, eg shuttering, formwork on or off site.

4.20 Timber works

- Off-cuts or damaged timber should only be disposed of if they cannot be reused or recycled.
- Working of timber should be carried out in a designated screened area:
  - to contain windblown dust that could be hazardous to humans and impact sensitive flora and fauna

  - to avoid pollution from preservatives
  - to avoid nuisance created by use of chainsaws and/ or planers
  - where site staff must wear appropriate PPE.
- Installation of timber piles, groynes etc involves tidal working and associated noise at anti-social hours:
  - if driving of piles is required, consider driving only on the day shift or using a muffled hammer or vibro-piling rig.

5 Post-construction phase

This phase of a construction project (ie at the end of the construction process) often receives less attention than the start-up and construction phases, but it can equally cause problems if it is not undertaken appropriately.

Normally, on completion of the works, the contractor is required to remove from the site all plant, surplus materials, rubbish and temporary works they are responsible for. The whole area of the site and works should be left clean – until then the project is not finished.

5.1 Site clearance/demobilisation

There will be conditions linked to the original permissions granted to allow the works to start that extend into the post-construction phase. These can include:

- landscaping
- removal of any debris or temporary works
- site reinstatement (ie structures removed temporally to allow construction).

5.2 Lessons learnt

During the delivery of a project there are often many lessons learnt such as following a pollution incident, which will be useful for future site management.

It is important that they are recorded, so any changes in practice can be embedded in the future and communicated to all site-based staff through site inductions, eg through toolbox talks (see Further guidance).

Embedding the learning not only improves practice, but can lead to time and cost savings as well as reducing the risk of damaging and organisations reputation.

5.3 Reporting

Following the construction of a project there may need to be some form of handover from the construction (and/or environmental) team to the operators of the site.

The types of environmental information that will need to be handed over may include:

- licences and consents, eg environmental permits
- records of compliance with licences and consents
- finalised design details, eg reports and drawings
- monitoring results – including any raw data and reports
- site management plans
- any documents recording lessons learnt (see **Section 5.2**)
- details of any consultation with the local community – including any issues during construction and how they were resolved.

5.4 Monitoring

Projects that have been predicted to have an impact on the environment will usually require monitoring to be continued post-construction.

The requirement for ongoing monitoring will form part of project consents for the project and/or will be contractual requirement. The site manager will be familiar with the requirements.

Post-construction monitoring can include measuring/counting environmental parameters, such as:

- physical processes, eg flows in navigation areas
- sediment quality
- water quality

- water birds
- marine ecology, eg invertebrates and fisheries
- vegetation
- habitat restoration/compensation schemes (and their success)
- the historic environment/archaeology, eg *in situ* stability following works.

A1 Useful contacts

The responsibilities of following consultees are as follows:

Marine regulator	Marine licensing, marine planning, oil spills at sea, monitoring and enforcement, marine wildlife and habitat conservation
Environmental agencies	Waste, discharges to controlled waters, abstraction licences, some nature conservation functions, flood defences and some response to spills
Statutory Nature Conservation Organisation	Designated ecological sites, geological and geomorphological sites, protected species and habitats
Environment department	Policy and regulations on the environment
Department for Communities and Local Government	Planning policy
Local authority	Noise, air quality, traffic, considerate constructors' schemes, ground contamination, Part B authorisations for air quality, footpaths and roads, landscape and aesthetics, designation of Local Nature Reserves (LNR), planning permission
County archaeologist, heritage bodies, landscape architects	Designated archaeological and heritage sites, landscape and seascapes, and aesthetics
Health and safety	Response to spills and marine safety
Port authority	Conservancy responsibilities enacted by Parliament

In addition, consult with water and sewerage undertakers regarding obtaining consents to discharge sewage and trade effluent (eg concrete wash water) to foul sewer.

England		
Marine regulator	Marine Management Organisation	T: 0300 123 1032 W: https://www.gov.uk/mmo
Environmental agency	Environment Agency	T: 03708 506 506 W: https://www.gov.uk/ea
Statutory Nature Conservation Organisation	Natural England	T: 0300 060 6000 W: https://www.gov.uk/natural-england
Environmental department	Department for Environment, Food and Rural Affairs	T: 0345 933 5577 W: https://www.gov.uk/defra
Local government	Department for Communities and Local Government	T: 0303 444 0000 W: https://www.gov.uk/dclg
Local authority	Various	W: www.gov.uk/find-your-local-council
Heritage body	Historic England	T: 020 7973 37 00 W: https://historicengland.org.uk

Wales		
Marine regulator	Natural Resources Wales (Marine Licensing Team)	T: 0300 065 3000 W: https://naturalresources.wales/?lang=en
Environmental agency	Natural Resources Wales	T: 0300 065 3000 W: https://naturalresources.wales/?lang=en
Statutory Nature Conservation Organisation	Natural Resources Wales	T: 0300 065 3000 W: https://naturalresources.wales/?lang=en
Environment department	Welsh Government	T: 0300 060 3300 W: http://gov.wales/?lang=en
Local government	Welsh Government	T: 0300 0603300 W: http://gov.wales/?lang=en
Local authority	Various	W: www.wales.gov.uk/topics/localgovernment
Heritage body	Cadw	T: 01443 336 000 W: http://cadw.gov.wales/?lang=en

Scotland		
Marine regulator	Marine Scotland	T: 0131 556 8400 W: www.gov.scot/marinescotland
Environmental agency	Scottish Environment Protection Agency	T: 03000 99 66 99 W: www.sepa.org.uk
Statutory Nature Conservation Organisation	Scottish Natural Heritage	T: 01463 725 000 W: www.snh.gov.uk
Environment department	Scottish Government	T: 0845 603 7602 W: www.gov.scot
Local government	Scottish Government	T: 0845 603 7602 W: www.gov.scot
Local authority	Various	W: www.cosla.gov.uk/scottish-local-government
Heritage body	Historic Scotland	W: www.historic-scotland.gov.uk T: 0845 603 7602

Northern Ireland		
Marine regulator	Department of the Environment NI	T: 028 9054 0540 W: www.doeni.gov.uk
Environmental agency	Northern Ireland Environment Agency	T: 0845 302 0008 W: www.nidirect.gov.uk/northern-ireland-environment-agency
Statutory Nature Conservation Organisation	Northern Ireland Environment Agency	T: 0845 302 0008 W: www.nidirect.gov.uk/northern-ireland-environment-agency
Environment department	Department of the Environment NI	T: 028 9054 0540 W: www.doeni.gov.uk
Local government	Northern Ireland Environment Agency	T: 0845 302 0008 W: www.nidirect.gov.uk/northern-ireland-environment-agency
Local authority	Various	W: www.nidirect.gov.uk/local-councils-in-northern-ireland
Heritage body	Northern Ireland Environment Agency	T: 0845 302 0008 W: www.nidirect.gov.uk/northern-ireland-environment-agency

Pan-UK		
Health and safety	Health and Safety Executive	T: 0300 003 1747 W: www.hse.gov.uk
Government agency	Maritime and Coastguard Agency	T: 023 8032 9100 W: https://www.gov.uk/mca
Port authority	UK Major Ports Group	T: 0845 302 0008 W: http://ukmajorports.org.uk

Further reading

CHARLES, P and EDWARDS, P (2015) *Environmental good practice on site guide, fourth edition*, C741, CIRIA, London (ISBN: 978-0-86017-746-3). Go to: **www.ciria.org**

COOPER, N and COOKE, S (2015) *Assessment and management of unexploded ordnance risk in the marine environment*, C754, CIRIA, London (ISBN: 978-0-86017-760-9). Go to: **www.ciria.org**

IMO (2012) *Guidelines for the control and management of ships' biofouling to minimise the transfer of invasive aquatic species, 2012 edition*, International Maritime Organisation, Florida, USA (ISBN: 9-789-28011-545-1).
Go to: **http://tinyurl.com/q2qvxcx** (accessed 22/02/2016)

JNCC (2004) *Common standards monitoring guidance for earth science sites*, Joint Nature Conservation Committee, Peterborough. Go to: **http://tinyurl.com/pcl8rkw** (accessed 22/02/2016)

JNCC (2004) *Guidelines for minimising the risk of disturbance and injury to marine mammals from seismic surveys*, Joint Nature Conservation Committee, Aberdeen.
Go to: **http://tinyurl.com/pokzrc3** (accessed 22/02/2016)

JNCC (2008) *The deliberate disturbance of marine European protected species. Guidance for English and Welsh territorial waters and the UK offshore marine area*, Joint Nature Conservation Committee, Aberdeen, Scotland.
Go to: **http://tinyurl.com/nqpz8xt** (accessed 22/02/2016)

JNCC (2010) *Guidelines for minimising the risk of injury to marine mammals from using explosives*, Joint Nature Conservation Committee, Aberdeen, Scotland.
Go to: **http://tinyurl.com/qxgmcwl** (accessed 22/02/2016)

JNCC (2010) *Handbook for Phase 1 habitat survey – a technique for environmental audit*, Joint nature Conservation Committee, Peterborough, UK (ISBN: 978-0-86139-636-8).
Go to: **http://jncc.defra.gov.uk/page-2468#download** (accessed 22/02/2016)

JNCC (2010) *Statutory nature conservation agency protocol for minimising the risk of injury to marine mammals from piling noise*, Joint Nature Conservation Committee, Aberdeen, Scotland. Go to: **http://tinyurl.com/qb94xkr** (accessed 22/02/2016)

JOHN, S, MEAKINS, N, BASFORD, K, CRAVEN, H and CHARLES, P (eds) (2015) *Coastal and marine environmental site guide (second edition)*, C744, CIRIA, London (ISBN: 978-0-86017-749-4). Go to: **www.ciria.org**

LAW, C and D'ALEO, S (2016) *Environmental good practice on site – pocket book (fourth edition)*, C762, CIRIA, London (ISBN: 978-0-86017-777-7). Go to: **www.ciria.org**

MARINE SCOTLAND (2014) *The protection of Marine European Protected Species from injury and disturbance Guidance for Scottish Inshore Waters*, Marine Scotland, Edinburgh. Go to: **www.gov.scot/resource/0044/00446679.pdf** (accessed 22/02/2016)

MURNANE, E, HEAP, A and SWAIN, A (2006) *Control of water pollution from linear construction project. Site guide*, C649, CIRIA, London (ISBN: 978-0-86017-649-7). Go to: **www.ciria.org**

PAYNE, R D, COOK, E J and MACLEOD, A (2014) *Marine biosecurity planning. Guidance for producing site and operation-based plans for preventing the introduction of non-native species*, Scottish Natural Heritage, Inverness, Scotland. Go to: **www.snh.gov.uk/docs/A1294630.pdf** (accessed 22/02/2016)

WADE, M. BOOY, O and WHITE, V (2008) *Invasive species management for infrastructure managers and the construction industry*, C679, CIRIA, London (ISBN: 978-0-86017-679-4). Go to: **www.ciria.org**

Toolbox talks

CIRIA has supported the updating of a suite of environmental toolbox talks that can be accessed from: **www.ciria.org/egpos**

Pollution Prevention Guidelines (PPGs)

Guidance published by the Environment Agency, SEPA and NIEA on specific aspects of pollution prevention including:

PPG 1 *Understanding your environmental responsibilities – good environmental practices*

PPG 5 *Works and maintenance in or near water*

PPG 6 *Working at construction and demolition sites*

PPG 7 *Safe operation of refuelling facilities*

PPG 21 *Incident response planning*

PPG 22 *Dealing with spills*

Copies of these PPGs and details of the all those available can be accessed from: **http://tinyurl.com/zjpn6km**

Glossary

Archaeology	Historic and prehistoric remains often discovered by excavation and found on construction sites.
Biodiversity	A term used to describe the variety of living things on earth. Maintaining or increasing biodiversity is a positive outcome.
Built environment	Developed areas, including residential, commercial and industrial property.
Consent	Agreement or permission to undertake an activity, which can include various permits and licences.
Conservation	A series of measures required to maintain or restore natural habitats and populations of wild flora and fauna.
Contaminated	Land that contains substances in, on or under land likely to represent a hazard to humans, animals or the environment.
Dust	Airborne solid matter up to about 2 mm in size. Along with noise and odour, dust is probably the most commonly complained about issue or 'statutory nuisance'.
Ecology	All living things, such as trees, flowering plants, insects, birds and mammals and their habitats.
Environmental Action Plan (EAP)	A report that ensures commitment to the adoption of mitigation and management measures, as recommended within an Environmental Statement (ES).

Environmental Impact Assessment (EIA)	A procedure employed to assess the likely significant impacts of a proposed development upon the environment.
Environmental Statement (ES)	The report or final set of documents containing the findings and recommendations of an EIA.
Fauna	The term used to describe the animal life of a particular place, region or period of time.
Flora	Term used to describe the plant life of a particular place, region or period of time.
Geomorphology	The study of the physical features of the Earth's surface and their formation (eg processes of erosion, transportation, and deposition).
Groundwater	Water beneath the ground's surface.
Habitat	The place where an organism lives, often defined based on uniformity of vegetation (woodland, reed bed etc).
Historic environment	All aspects of the environment resulting from the interaction between people and places through time, including all surviving physical remains of past human activity, whether visible, buried or submerged, and landscaped and planted or managed flora.
Intertidal	The zone between high and low tide marks that varies in extent along the coast.
Invasive species	Plants or animals that have the ability to spread causing damage to the environment, economy, health and the way people live.

Maintenance	The general task of ensuring that construction plant and (plant and vehicles) vehicles are kept 'roadworthy' and capable of undertaking their allotted tasks.
Mean high water springs (MHWS)	The average of the heights of two successive high waters during those periods of 24 hours (about once a fortnight) when the range of the tide is greatest.
Mitigation	Measures taken to reduce adverse effects.
Natural environment	The composition of wildlife and habitats.
Noise	A sound that is not desired. Sound is a wave motion carried by air particles between the source and the receiver (usually the ear).
Offshore	That area of sea beyond the land with depths in excess of 20 m.
Organism	Any single living plant or animal
Pollution	Introducing substances or energy (eg noise) to land, water or air that are likely to cause significant harm to humans, wildlife and/or buildings. Pollutants include silty water, heavy metals, sewage, oils, chemicals, litter, clays and mud.
Protected species	Certain plant or animal species that are protected to various degrees in law.
Recycling	Collecting and separating materials from waste and processing them to produce marketable products.
Reuse	Putting objects back into use, without processing, so that they do not remain in the waste stream.

Risk	The chance of an adverse event actually occurring.
Silt	Waterborne particles with a very small grain size.
Species	A group of organisms that interbreed.
Subtidal	The region of the sea and the seabed that occurs beneath the tidal zone.
Tide	The periodic rise and fall in the level of water in the oceans and sea, resulting mainly from the gravitational attraction of the sun and the moon.
UK waters	This includes territorial waters, where limits extend to 12 nautical miles offshore, and waters to 200 nautical miles offshore within which the UK exercises certain rights and jurisdictions.
Waste	Any substance or object that the holder discards, intends to discard, or is required to discard:

- controlled waste – household, commercial, industrial
- directive waste – material that the producer or holder discards
- inert waste - material that does not undergo any significant physical, chemical or biological transformations. Inert waste will not dissolve, burn or otherwise physically or chemically react, eg clean bricks or concrete
- non-hazardous waste – material that does not have any significant hazardous properties, but is not inert and could

cause problems if not dealt with properly because it may biodegrade, eg paper, cardboard or plastic

- hazardous/special waste – can be harmful to the environment and human health and so cannot be disposed of by conventional methods, eg paints, solvents, oil and pesticides.

Wildlife Any undomesticated organism.

Acronyms and abbreviations

BPM	Best practicable means
COSHH	Control of Substances Hazardous to Health
DCLG	Department for Communities and Local Government
Defra	Department for Environment Food and Rural Affairs
DOENI	Department of the Environment Northern Ireland
EAP	Environmental Action Plan
EIA	Environmental Impact Assessment
EMP	Environmental Management Plan
ES	Environmental Statement
HSE	Health and Safety Executive
IMO	International Maritime Organisation
INNS	Invasive non-native species
JNCC	Joint Nature Conservation Committee
LNR	Local Nature Reserve
MarLIN	Marine Life Information Network
MCA	Maritime and Coastguard Agency
MHWS	Mean high water springs
MMO	Marine Management Organisation
MMMP	Marine Mammal Mitigation Protocol
NIEA	Northern Ireland Environment Agency
NNSS	Non-native species secretariat
NRW	Natural Resources Wales
PPE	Personal protective equipment
PPGs	Pollution Prevention Guidance notes
RMP	Resource Management Plan

SEPA	Scottish Environment Protection Agency
SNCO	Statutory Nature Conservation Organisation
SNH	Scottish Natural Heritage
SWMP	Site Waste Management Plan
UKMPG	United Kingdom Major Ports Group
UXO	Unexploded ordnance
VOC	Volatile organic compounds
WFD	Water Framework Directive
WG	Welsh Government
WRAP	Waste & Resources Action Programme

Acknowledgements

This pocket book updates the original guide (CIRIA C594) published 2003. It has been undertaken in collaboration with a project steering group (PSG) of industry practitioners.

CIRIA would like to acknowledge those involved in the development of the original pocket book in 2003 (C594).

Editors

Sian John	Royal HaskoningDHV
Katharine Basford	Royal HaskoningDHV
Helen Craven	Royal HaskoningDHV
Dr Nicola Meakins	Southern Water (formally Kasa Consulting UK Ltd™)
Philip Charles	CIRIA
Sirio D'Aleo	CIRIA

Project steering group

CIRIA wishes to express its thanks to the members of the group for their significant contribution to the pocket book:

Rebecca Clough (chair)	Scottish and Southern Energy
James Addicott	Eastern Solent Coastal Partnership
Siobhan Browne	Natural England
Stuart Churchley	Historic England
Phil Elliott	Northern Ireland Environment Agency
Lucinda Farrington	Seriously Green
Gavin Holder	Eastern Solent Coastal Partnership
Dr Liz Howe	Natural Resources Wales
Mark Johnston	Natural England

Nigel Johnston	BAM Nuttall
Patrick Jordan	Aberdeen Harbour Board
Daniel Leggett	DLEnviro
Tom Matthewson	HR Wallingford
Neil Martin	Boskalis Westminster Ltd
Richard Robinson	Lagan Construction Group
Andrew Wallace	Scottish Environment Protection Agency
Steve Wenham	Environment Agency
Tim Yzewyn	DEME nv

Source materials

CIRIA would like to acknowledge the following organisations for providing figures (diagrams and photographs):

Aberystwyth University, BAM Nuttall Ltd, DEME nv, East Solent Coastal Partnership, Green Construction Board, G Skomal, London Gateway Port, Paul Beckwith BWW, Royal HaskoningDHV, R Saunders, VolkerStevin Ltd, Waste & Resources Action Programme, Wessex Archaeology

CIRIA project team

Philip Charles	Project manager
Sirio D'Aleo	Assistant project manager

Funders

BAM Nuttall Ltd

DEME nv

Scottish Environment Protection Agency

Scottish and Southern Energy

CIRIA Core member programme